DIPLALO TSA DIPALE

THE NUMBER STORY

SMALL BOOK ONE
ENGLISH - SESOTHO

Numbers Teach Children
Their Number Names

written and illustrated by

MISS ANNA

Early Reader Edition of *The Number Story 1*
Bronze Medal Winner, 2016 Wishing Shelf Book Award

Library of Congress Control Number: 2018902040

Names: Miss Anna, author.
Title: Number story : numbers teach children their number names / Miss Anna.
Description: Portland, OR: Lumpy Publishing, 2018.
Identifiers: ISBN 978-1-949320-20-6| LCCN 2018902040
Summary: The pictures and rhymes present stories which introduce numbers 0-10.
Subjects: LCSH Numeration—English—Sesotho--Pictorial works--Juvenile literature. | BISAC JUVENILE NONFICTION /
Languages: English—Sesotho
Classification: LCC QA141.3 .M57 2018 | DDC 513—dc23

Publisher: Lumpy Publishing
Website: www.missannabooks.com
Email: missanna@missannabooks.com

Paperback: ISBN 978-1-949320-20-6
Printed in the U.S.A.   1 3 5 7 9 10 8 6 4 2

Ana oka rata hoithuta mabitso a dipalo?

It is very easy and a lot of fun!

E bobebe haholo ebile eya thabisa!

Say-along our little jingle

Bina le rona dipalenyana tsa rona!

starting from Number One!

Retla qala ho Nngwe!

1

ONE   looks like my one finger.

NNGWE

etshwana le monwana
waka ole mong.

ONE!
NNGWE!

2

TWO   trails a tail.

PEDI

e hula mohatla.

A TAIL! MOHATLA!

3

THREE   has bumps.

THARO

e na le bampara.

BUMPY! BAMPARA!

4

FOUR carries a sail.

NNE

e na le sekepe.

4
A SAIL!
SEKEPE!

5

FIVE   is a racing track.

HLANO

ke tsela e sebelisetoang ho palama.

VROOM
VRUUM!

6

A SNAIL! SEMAMINA!

7

SEVEN has a sharp angle.

SUPA

ena le kgutlatharo e mutsu.

BE CAREFUL! IT'S SHARP!

EBA LE HLOHKO! EMUTSU!

8

EIGHT   is rollercoaster rails.

ROBEDI

ke tsela ea rola kowsta.

YEY!
YIPPEE!

9

NINE is a bubble on a stick.

ROBONG

dibudulwana hodima thupa.

A BUBBLE!

SEBUDULWANA!

10

TEN is an eye of a whale.

LESHOME

ke leihlo le le leng la leruwaruwa.

WINK!

PANYA!

HELLO! DUMELA!

And
Hape

0

ZERO   is an empty pail.

LETHO

ke nkgo esenang letho.

IT'S
EMPTY!

HAHOLETHO!

Thank you for playing with us today.

We had a lot of fun too!

Re a leboha o bapetse le rona kajeno.

Re bile le nako e monate haholo!

We are your Number friends,
Zero to Ten,
Who will be here for you~
Re batswalle ba hao ba dinomoro
Noto ho isa ho Leshome.
Retla dula leteng bakeng sa hao.

Bye-bye now!
See you again soon!
Sala hantle!
Retla bonana hape ka
nako esa fedising pelo!

The Numbers are *SINGING* too!

To sing-a-long, look for Miss Anna Number Story
at your favorite music store like iTUNES.

MP3

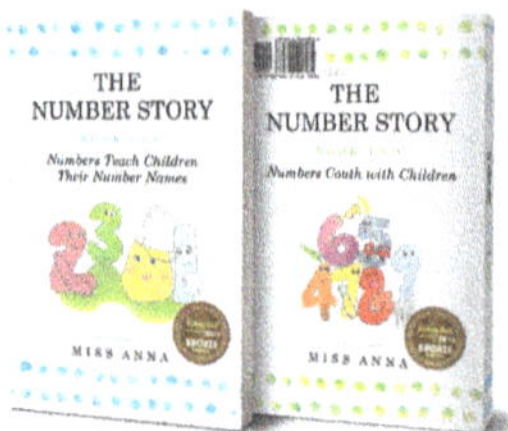

Numbers 0-10
IDENTIFYING
& COUNTING

Number Story 1 & 2
isbn: 978-0-996216-48-7

Numbers 11-20
& Ordinals
first, second, third...

Number Story 3 & 4
isbn: 978-1-945977-01-5

Numbers 0-100
& Place Values
ones, tens, hundreds...

Number Story 5 & 6
isbn: 978-1-945977-06-0

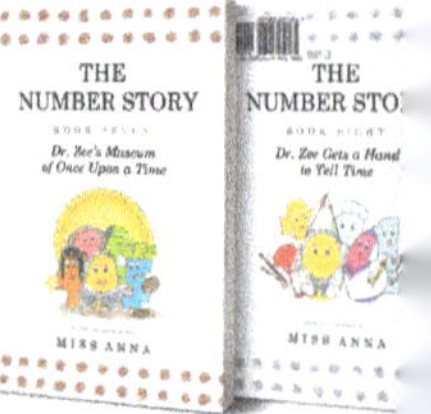

About Clock
& Telling Tir
hours, minutes, seco

Number Story 7 &
isbn: 978-1-949320-

For more Miss Anna books to love,
visit us at

w w w . m i s s a n n a b o o k s . c o m

Numbers are working hard all over the world!
Come Travel the World with Us!